INTERCITY HST 125

Hugh Llewelyn

AMBERLEY

First published 2014

Amberley Publishing
The Hill, Stroud
Gloucestershire, GL5 4EP

www.amberley-books.com

Copyright © Hugh Llewelyn, 2014

The right of Hugh Llewelyn to be identified as the Author of this work has been
asserted in accordance with the Copyrights, Designs and Patents Act 1988.

ISBN 978 1 4456 3418 0 (PRINT)
ISBN 978 1 4456 3425 8 (EBOOK)

British Library Cataloguing in Publication Data.
A catalogue record for this book is available from the British Library.

Typeset in 11pt on 12pt Sabon LT Std.
Typesetting by Amberley Publishing.
Printed in the UK.

CONTENTS

FOREWORD

This book is a not an illustrated history of the high speed train; that has been done admirably by Colin Marsden in his two volumes *HST Silver Jubilee* and *HST The Second Millennium*. Neither is it a photographic record of the many liveries carried by HSTs both under BR and under privatisation; that has been done comprehensively by Gavin Morrison in *The Heyday of the HST*. This book is simply a selection of the many photographs I have taken over the years since I first saw the prototype HST in Paddington, not long after it had entered revenue service in 1975.

There are gaps both geographically and in timescale in my photographs. Bringing up a family and having a busy career meant that my photography in the 1980s and 1990s was relatively limited, and the fact that, from 1976 onwards, I have lived in the Bristol area has meant that my photographs are heavily biased towards the Paddington–Bristol main line and, like many photographers, I have my favourite vantage spots. However, a bias towards the Western Region/Great Western Main Line does not necessarily present a false impression of HST deployment. The Western Region of British Railways was the first to introduce the HST in revenue service (the prototype), and was the first to introduce large-scale HST services. The Great Western Main Line was also where the HST provided – and will still provide for a few more years yet – the mainstay of express passenger services. In contrast, the HST now plays second fiddle to Class 91s on the East Coast Main Line, to Meridians on the Midland Main Line and to Voyagers on Cross Country routes.

So, though arguably uneven in coverage, I hope that my pictures and accompanying captions bring some pleasure to the reader.

I would like to give a word of explanation about the captions to my photographs and the identification of the locomotives pictured. Under BR auspices, all HST locomotives were classified Class 43 and had Paxman Valenta 12RP200 engines (apart from some examples temporarily fitted with Mirlees engines). The pictures I took in this era are thus merely identified as 'Class 43'. However, under privatisation, all the Valentas were gradually replaced by Paxman 12VP185 and MTU 16V4000R14R engines, and the locomotives became Class 43/0 or 43/2 – not necessarily coinciding with the make of engines fitted! My captions of this era identify both the subclass and engine fitted.

INTRODUCTION

I have long been a railway enthusiast. I grew up in the South Wales valleys with the old Neath & Brecon Railway at the bottom of my garden. Coal trains and two-coach passenger trains hauled by the ubiquitous ex-Great Western pannier tanks were my staple diet, but railways were only a mild interest until in 1961 I went to Neath Grammar School. Its location was heaven for a young 'trainspotter'. At the bottom of the playing fields was the main Paddington–Swansea main line. My interest in trains was ignited, seeing close at hand expresses hauled by gleaming Castles and Halls – well, that's how I like to think they looked through the soft-focus lens of nostalgia. In reality most steam locomotives in those days ranged from grubby to absolutely filthy, and the coaches were often only a little better.

I thought the days of steam and smut in my eyes were going to last forever, and the first Hymek and Western diesel hydraulics I saw did little to alter that. My friends worshipped at the altar of GWR steam and could not understand how any self-respecting trainspotter could be interested in BR Standard steam locos, let alone diesels. They could not begin to understand how I, in marked contrast to them, was interested in all steam locomotives whatever their origin and – almost unbelievably – even in modern traction. While I naïvely thought steam would last for many years, I assumed the Westerns and Hymeks would last for decades. Little did I know that by 1968 steam would have gone from the entire British Railways network, that all bar one of the various diesel-hydraulic classes would be withdrawn by the mid-1970s and that even the impressive Westerns would last for only a few years more than their sister diesel hydraulics. In fact, as the final Westerns disappeared to the scrapyard (or into preservation), a 'new kid on the block' arrived on the scene, a train that would have been hardly imaginable to a young trainspotter ten years previously – the High Speed Train.

My first encounter with a High Speed Train was in June 1975. I was at that time a civil servant working in Westminster in the Railway Directorate of the Department of the Environment – there was no separate Department of Transport in those days! I don't remember exactly why I was to travel on the prototype HST – or HSDT (High Speed Diesel Train) as it was then known – but I do remember what it was like.

Standing at the end of the platform at Paddington that June

morning, my immediate impression was how long the HST coaches were. At 23 metres, the Mark 3 coaches were considerably longer than the then current generation of coaches, the 19.6-metre Mark 2 coaches. The coaches were also designed to be easily converted to/from either class and to have various seating densities and layouts. This has indeed happened several times during the various refurbishment programmes that the Mark 3s have been subjected to over the years. Even the catering vehicles have been subject to considerable change in layout and role, and the original three types of catering coaches later multiplied to many more.

Coupled with the streamlined fairings that enclosed the variety of equipment under the coach so neatly, the Mark 3s certainly looked modern and, well, fast. I wasn't so sure about the aesthetics of the HST prototype locomotive, however. With a length of 17.2 metres, I thought the loco looked a little too short to match the coaches. Having always been impressed by the 20.7-metre length of the Peak Class diesels I was used to seeing at Cardiff and Bristol, I saw the HST locos as somewhat stunted. The rather blunt nose of the loco with a small windscreen and no side windows in the cab seemed to add to that compressed look, which was further emphasised by the large expanse of the mandatory yellow warning panel that enveloped the entire cab and wrapped around its side.

Apart from the yellow cab end, I was very impressed by the livery of the prototype train. It copied the final livery of the prematurely withdrawn Metro-Cammell Blue Pullman DMUs. That livery was the reverse of the standard InterCity livery: very light grey (more an off-white) with a blue panel along the windows, continuing along the locomotives' flanks. It had a touch of 'class' about it.

The interior of the Mark 3 coaches was not quite the revelation I'd anticipated. True, they were a vast improvement on the Mark 1 and early Mark 2 coaches that predominated on expresses, but I didn't think they were very much better than the latest air-conditioned Mark 2 coaches. Those were very comfortable and quiet, and the Mark 3s simply continued that environment. There was one aspect of the Mark 3s that was a little disappointing, however, and that was the smell of hot brake dust under heavy braking, a characteristic that took years to rectify in the production sets.

Nonetheless, overall, my impressions of my first HST journey to Swindon were very favourable. Over the years those impressions have not changed. I have since travelled frequently on HSTs both on business and for pleasure, mainly on Paddington–Bristol/South Wales services but also occasionally on Cross Country and sometimes ECML services.

It was not just the fact that the HST was designed for routine and prolonged high speed that was a 'game changer' for BR, but the very concept of the train. Up until then, virtually all long-distance express services consisted of individual coaches hauled by a locomotive – exactly the same as in the days of steam. In times of high demand, a coach or two or more could be hung on the end of a particular train. That was to change with the introduction of the HST. The prototype copied the concept of the multiple unit, which had been built in very large numbers for local, suburban and short-distance express services since the mid-1950s. There was one particular design of DMU that in many ways was the precursor of the HST. Those were the ground-breaking Metro-Cammell Blue Pullman 6 and 8 car DMUs of 1960, mentioned above. Their concept was very similar to the HSTs, with semi-permanently coupled coaches marshalled between two power cars. Although the Blue Pullmans differed from the HSTs in having some passenger accommodation in the power cars, the big difference was that the Metro-Cammell trains were for specialist Pullman services catering for prosperous businessmen who could afford the extra premium required to travel on them – the HSTs were intended to bring luxury and high speed to the 'masses' at standard fares!

The prototype HST was thus a fixed, semi-permanent formation of seven coaches (or trailers) with two locomotives (or power cars), one at either end. This greatly facilitated turnaround times at terminals. No more would a locomotive at one end have to be detached at the finish of each journey and a fresh locomotive attached to the other end for the return journey. With quicker turnaround times and much faster journeys, an enormous saving in rolling stock was anticipated with the squadron introduction of HSTs, all to the benefit of BR's precarious finances.

Little did I realise on that June day in 1975 that the prototype train I was gazing at under Brunel's roof at Paddington would presage the demise of so many loco-hauled trains in Britain, to be replaced by HSTs, Pendolinos, Adelantes, Voyagers, Meridians and such like. That seems to me almost as significant a change as the end of steam.

However, the HST has always been a bit schizophrenic – it has been undecided whether it is a loco-hauled train or a multiple unit! Like so many British engineering successes, it was conceived merely as a 'stopgap'. In the late 1960s, British Rail saw its future as widespread electrification and a large fleet of Advanced Passenger Trains, then in the very early design stages. It was even envisaged that, for non-electrified lines, gas turbine APTs would be built, though that quickly died a death when it was realised how expensive they would be to run. Designing the electric APT also ran into problems very early on.

The APT project was authorised in 1968, but in the following year it was realised that there would be considerable delays in introducing the fleet, and a serious reliability problem with the tilting mechanism for the coaches was emerging. Moreover, the chances of future electrification seemed to be lessening. However, increasing competition from aircraft and the motorways meant there was an urgent need for new and improved trains. Already on the drawing boards at Derby was a new design of locomotive-hauled coach to replace the Mark 2, to be designated the Mark 3. What was needed was a very high-powered locomotive to haul the Mark 3 coaches at the 125 mph deemed necessary to combat the competition. However, as no single lightweight diesel engine in the 4,500 hp class was in existence, the solution was to place a locomotive of half that power at either end of the train. This would also have the advantage of shortened turnaround times at terminals. A suitable engine already existed – the Paxman Valenta, with 2,250 hp.

So, the HSDT was born. The project, led by Walter Jowatt, Director of Design, and Terry Miller, Chief Engineer of Traction and Rolling Stock, was authorised in 1970. The prototype was not at first regarded as a DMU. The locomotives were classed as such (Class 41 and numbered 41 001 and 41 002) and the coaches were numbered in the loco-hauled series 1XXXX. It was thus when the prototype emerged in 1972–73. The coaches were built by BREL at Derby, the locomotives at Crewe. It is notable that the locomotives had buffers and draw gear at the No. 1 end and a second, rather basic, slab-ended cab at the other end, with controls suitable for short-term use and shunting. As locomotives, it was expected that they would haul freight trains (coupled back-to-back), high-speed container trains and sleeper trains. Adjacent to the tiny rear cab was a compartment for the conductor (i.e. guard) and luggage area. Trials on the Eastern Region went well and in May 1973, 252 001 set a world speed record for diesel trains of 143.2 mph.

After the completion of the ER trials and an overhaul at Derby Works in May 1974, thoughts on how the HSTs would be employed changed. They were in future to be regarded as multiple units, with the locomotives now 'power cars' (or strictly speaking driving motor brakes – DMBs) and the coaches 'trailers'; all were renumbered in a new 4XXXX series particular to HST stock, the power cars being numbered 43000 and 43001. The prototype 'unit' – designated the sole Class 252 DMU and numbered 252 001 - was transferred to the Western Region in December 1974. After further testing, the prototype began in passenger service in May 1975. It was one of those early services that I caught – the 10.15 Paddington–Weston-super-Mare, although I travelled only as far as Swindon.

The prototype consisted of seven trailers between the power cars with, surprising to us these days, two catering vehicles – a kitchen/restaurant car and a buffet car. The early production sets – classed 253 – for the Western Region initially mirrored this formation, though very quickly it was realised that the provision of two catering cars was excessive bearing in mind rapidly changing demands for on-train catering. Thus, one catering vehicle was replaced by a saloon.

The production HSTs differed little from the prototype excepting the power cars. No longer being regarded as locomotives, there would be no use for an inner cab, and that was replaced by an enlarged luggage and guard's compartment, while the buffers at the No. 1 end were removed and replaced by a streamlined fairing. Interestingly, after the production of the early sets, trade union objections to the guard's accommodation next to the noisy engine compartment and to the rough ride of the power car resulted in that accommodation being moved to a trailer, which also had some luggage accommodation provided. Although the first 151 production Class 43s were, like the prototypes, classed as DMBs and had the same glazing arrangements at the rear end, the last forty-six locomotives built had the rear end window and the two rear side windows removed; they were classified simply as DMs – driving motors. The glazing of the earlier locomotives was gradually, over many years, altered to match.

In addition, the union had objected to the prototype locomotive's single driving position in the main cab with its limited glazing; as a result, the production locos had a completely revised cab (by industrial designer Kenneth Granger) for a crew of two, an enlarged windscreen and side windows. Aesthetically, the result was a greatly improved 'front end'.

However, that improvement was negated by the livery chosen, in my opinion. Leaving aside the hideous black and yellow livery of the first production power car completed in 1975 (which was mercifully repainted in the new scheme finally chosen before entering service), the Class 253s reverted to the standard InterCity livery of a blue base with light grey window surrounds instead of the very attractive 'reversed' scheme of the prototype. Presumably someone thought that that livery would be too difficult to keep clean. The standard InterCity scheme was reasonably attractive on the coaches – sorry, trailers – but the treatment of the power cars did not really work, in my opinion. Continuing the blue/grey scheme to the guard's compartment of the power cars but then raising the blue base above the grill to be carried over a large expanse of warning yellow stretching from the cab resulted in a very disjointed appearance, to the detriment of what was otherwise a good-looking power car – though still rather short!

The introduction of an HST service to the WR in 1976 resulted in the fastest diesel-operated passenger service in the world. HST services were subsequently phased into service on the East Coast Main Line (with eight trailers and classified '254') in 1977–79, and on Cross Country routes (with seven trailers and thus Class '253') in 1981–82. As originally envisaged, British Rail proposed a fleet of 161 sets for use on the Great Western Main Line, the East Coast Main Line, Cross Country routes, the Midland Main Line and even Trans-Pennine and Glasgow–Edinburgh services. However, after the delivery of what was thought to be at the time just the first tranche of sets for Cross Country, those plans went awry. The UK's declining financial situation in the 1980s, coupled with a drop in demand for long-distance rail travel, saw the cancellation of the HST programme after only ninety-five sets were delivered. Nonetheless, with judicious juggling of resources and the construction of some additional trailers in 1982, BR managed to reallocate sets from existing routes to the Midland with the result that HST services (with eight trailers but classified 253 rather than the more logical 254!) were introduced out of St Pancras in 1983.

It is amazing to think that the production HSTs were anticipated to have a main-line life of just ten years, by which time they would be worn out! They would then be switched to less demanding secondary services. By then, widespread electrification and the production of APTs would be well underway – or so it was thought. In the event, HSTs have had to continue working at full throttle well past their 'sell-by-date'. Despite the electrification of the ECML in 1988–90 displacing many sets (but far from all) on that route, the privatisation of the network and growing passenger numbers saw an increased demand for HSTs. For example, in 2006 the open access operator Grand Central wished to operate an HST service (with five or six trailers in each set) from King's Cross to Sunderland, but to obtain the necessary stock, GC had to rebuild long-stored power cars, which were little more than wrecks, while many of the trailers needed were converted from standard, non-HST loco-hauled coaches. That service commenced in late 2007.

However, that's jumping ahead. With the 'sectorisation' of BR in the 1980s came another change in classification for the HST. Although the sets were formed in semi-permanent formations, the power cars were often juggled between sets because of failures and operational needs. Thus it was possible to see a set with a power car with a particular set number at one end, while the second power car would have a different number – and in between them were a set of trailers that might belong to one or other of the power cars but occasionally might not belong to either! Consequently, set numbers were often painted out on power cars. Not long after the introduction of the attractive 'InterCity Executive' livery in 1983, it was decided that HSTs should no longer be regarded as multiple units and to reclassify the power cars as Class 43 locomotives. The confusion of locomotives with both set numbers and car numbers had gone. However, even though the coaches were now technically hauled/

pushed by locomotives, they were still officially classified as 'trailers' and were allocated to semi-permanent sets (minus locomotives) with a new alphanumeric identification system!

In 1987, with the InterCity sector gaining greater independence, came what I regard as the most attractive livery ever carried by HSTs – the 'InterCity Swallow' scheme. A refinement of the 'InterCity Executive' livery, the coaches were much the same but the aesthetics of the locomotives were greatly improved by the confinement of the yellow warning area to the lower nose and cab roof, and the continuation of the coach livery over the entire locomotive. Privatisation of the railway industry in 1996 has resulted in the HSTs carrying many liveries over the years, not only as franchise holders changed, but even by single franchisees when it has been decided to change an image, presumably for marketing reasons. First Great Western is a prime example, having used three very different liveries.

An interesting development affecting eight HST locomotives occurred in 1987–88, which in some respects was a reversion to the look of the prototypes. With the electrification of the ECML, it had been planned to operate Class 91 electric locomotives and newly designed Mk 4 coaches, which would include a DVT (driving van trailer) at the other end of each train to enable push-pull operations. Unfortunately, delays in ordering the coaches meant the new, very expensive Class 91s were completed without any Mk 4 coaches to haul! The decision was taken to use Mk 3 loco-hauled coaches and convert eight Class 43 locomotives to 'surrogate DVTs'. Initially it was intended that only their driving positions should be in use, but in the event their engines were used as well. However, to act as a DVT meant the fitting not only of time division multiplex equipment but also buffers and a cutaway lower cab front – rather reminiscent of the prototype Class 41 locos. After the introduction of the Mk 4 coaches, the TDM equipment was removed, and the eight 'surrogate' DVTs reverted to standard Class 43 configuration – though retaining the buffers. Six of these eight locos are now operated by Grand Central and two by Network Rail.

Another change in appearance that has occurred to the locomotives over the years has been the replacement of the original light clusters flanking the warning horn grill at the front of the cab. Originally there were two sets of triple marker lights, each cluster behind a common glazed screen. Over the years, these have been replaced by individual triple light clusters on Midland locomotives and twin light clusters on all others, both without a covering screen. The shape of the grill was revised at the same time.

Over the years, the Great Western and Midland sets have been expanded to eight trailers and the ECML sets to nine. There have been several refurbishment programmes for the coaches, both under BR and the succeeding private companies. The locomotives have also been refurbished. All were originally powered by the Paxman Valenta 12RP200L with 2,250 hp; it had been intended to fit an upgraded version of 2,500 hp to the power cars of the eight trailer Class 254 sets, but fears about unreliability saw the standard engines fitted. Interestingly, Paxman at one stage proposed a 3,000 hp version of the Valenta, but nothing came of that. The performance of an HST with a 3,000 hp locomotive either end would have been mouth-watering – but it was not to be. Perhaps its unreliability would have been as spectacular as the performance!

By the early 1980s, BR wanted a more up-to-date, more reliable, more efficient engine. In 1986–87 four Class 43s were fitted with 2,400 hp Mirlees 12MB190 engines, and these apparently proved successful. It was proposed to re-engine almost all the fleet. However, with each BR sector becoming more independent financially, and the need to ensure best value for money, the re-engining programme went out to tender, and the engine chosen to re-equip the fleet was the successor to the Valenta – the 2,100 hp Paxman 12VP185. Following a trial fitting to one loco in 1994, the re-engining programme began in 1995, but things were changing rapidly in the rail industry. Under privatisation, while the new operating company Midland Mainline favoured the 12VP185 and continued the re-engining programme, the other companies operating HSTs had other ideas. Following Brush fitting the highly efficient, low-emission German 2,250 hp MTU 16V4000 R14R to two Class 43s owned by lessor Angel Trains in 2004–5, First Great Western trialled the cars in operational use and were greatly impressed by the MTU. In 2006 FGW commenced re-engining their entire fleet. GNER and its successor National Express East Coast soon followed. Later, Cross Country and Grand Central also adopted the MTU, as did Network Rail for its rail measurement set.

One interesting experiment in regard to the power train must be mentioned. That was the conversion of one Porterbook-owned Class 43 locomotive in 2006–07 to a hybrid technology testbed with dual diesel (Valenta-engined) and battery power, in conjunction with Hitachi. After a year of trails, the experiment came to an end and nothing more seems to have come out of it. The locomotive was then refurbished for East Midlands Trains with a Paxman VP185 engine.

The HSTs must be regarded as one of Britain's most successful railway products. Aside from the large increase in passenger numbers generated by the greatly improved services that resulted from the introduction of the sets, two world speed records are still held by HSTs. A shortened (five trailers) Class 254 set holds the world speed record of 144 mph for a diesel-hauled train with passengers, achieved on 27 September 1985 during a press run for the launch of the resurrected *Tees-Tyne Pullman*, while another set holds the outright record for a diesel train of 148 mph during a test run on 1 November 1987.

With the introduction of a full HST service on the WR in 1976, the prototype set was withdrawn. One of its power cars, No. 41 001, is preserved and until recently has been on display at the National Railway Museum, York, although currently it is on loan to the 125 Group and is out and about making 'guest appearances' at various railway galas. Meanwhile, the production sets are still hard at work and have to soldier on until the Hitachi Super Express trains come on stream in 2017/18, and even then some HSTs are expected to remain in service until the mid-2020s at least – so much for the HST's original life expectancy of ten years!

On 13 August 1975, Royal Mail issued a set of four stamps to commemorate the '150th Anniversary of Public Railways'. The 12p stamp showed an HST power car in black and yellow livery, which at that time was the intended livery.

High-Speed Train 12P

1975 British Rail Inter-City Service HST

HSTS IN THE BR REGIONS –
THE BLUE AND GREY ERA

The prototype of the High Speed Diesel Train, as it was then called, at Paddington on a Weston-super-Mare service, June 1975. I was just about to use this service to Swindon. The prototype train had been first classified as locomotive-hauled coaches with a Class 41 locomotive either side. However, before entering service the prototype had been re-designated as a Class 252 diesel multiple unit, consisting of two power cars plus seven trailers. The set was numbered No. 252 001 and the 2,250 hp Bo-Bo power cars were renumbered in the DMU car series 43XXX so were not technically classed as Class 43 locomotives. The illustration shows PC (power car) No. W43000 (ex-No. 41 001) leading in set No. 252 001, which is in original 'reverse InterCity' 'rail grey' and 'rail blue' livery. Note that the 43000 number confirms it as DMU car stock since cars started with an XXXX0 number while locos began with XXXX1. Also, DMU stock did not have a space between the first two numbers and the following three as locomotive numbers did. When classed as locomotives, the two Class 41s' numbers had no regional prefix, although the Mark 3 coaches did – 'E' for Eastern Region, where the HST was being tested. On reclassification to a DMU and transfer to the Western Region, the trailers acquired the 'W' prefix, as did the power cars, in line with DMU practice. In May 1973, No. 252 001 set a world speed record for diesel trains of 143.2 mph (230.5 kph).

PC (power car) No. W43001 (ex-No. 41 002) is at the rear of the prototype HSDT set No. 252 001 departing Swindon on a Paddington–Weston-super-Mare service, June 1975. I had just alighted from this train.

Initially classed as a DMU, HST set No. 253 011 in standard BR InterCity 125 'rail blue' and 'rail grey' livery at Bristol Parkway on a Paddington–South Wales service, June 1977. At the head of the set is power car No.W43022. Note that at the rear, the former guard's compartment of the power car still has a window behind the access door. The crowds are waiting for the passing of GWR Castle Class 4-6-0 No. 4079 *Pendennis Castle* on its way to Avonmouth Docks for Australia. The Western Region made a very good case for their fleet and all that they asked for was authorised by the government: twenty-seven sets for the South Wales/Bristol service (Nos 253 001–027 plus five spare locomotives) and fourteen sets for the West of England service – although in the event only thirteen sets for the latter service were formed (Nos 153 028–40), the other vehicles being used as spares.

At that time classed as a DMU (two power cars plus eight coaches), HST set No. 254 031 (without set number painted on front) in BR InterCity 125 'rail blue' and 'rail grey' livery, with power car (later locomotive Class 43) No. E43116 (with number in white ahead of cab door) leading, approaching York on a King's Cross-bound service, September 1981. What had been intended as the guard's window to the rear of the access window has not yet been plated over. The power car was later named *City of Kingston upon Hull* and subsequently *The Black Dyke Band*. HST power cars carried the regional prefix ahead of the number because they were regarded as simply a constituent vehicle of a DMU and all DMU's then had the regional prefix. 'Rail blue' was a mid-blue colour, while 'rail grey' was a very pale grey, more an off-white. Originally, the British Railways Board had ordered forty-two sets for the Eastern and Scottish Region's ECML operations, but because of deteriorating government finances, only thirty-two sets were authorised (Nos 254 001–032). However, higher passenger numbers than expected for the new services resulted in another five sets being completed later (Nos 254 033–037). Interestingly, because the ECML sets were to have one more car than the Western Region's, it had originally been proposed to uprate the Paxman Valenta engines to 2,500 hp, but in the event the power cars of the Class 253 sets remained at 2,250 hp for the sake of standardisation and inter-operability.

Class 253 HST No. 253 016 in BR InterCity 125 'rail blue' and 'rail grey' livery, with power car No. 43 032 (later named *The Royal Regiment of Wales*) in the rear passing Chelvey and approaching Nailsea & Backwell on a Weston-super-Mare–Paddington express, August 1978. The set consists of two power cars and seven trailers, the standard formation for Western Region sets when introduced. On blue and grey liveried power cars, the set numbers were in black and positioned on the front coupling cover.

Class 253 HST DMU (as then classified) set No. 253 017 in BR InterCity 125 'rail blue' and 'rail grey' livery, with power car No. W43034 (later named the *Black Horse* and then *Travel Watch South West*) leading (with set number in black on the coupling cover and power car number in white on the guard's compartment), arriving at Paddington passing similar liveried set No. 253 019, with power car No. W43039 (later named *The Royal Dragoon Guards*) leading, departing Paddington, September 1982. The Western Region was the only region to have their planned complement of forty-one HSTs fully approved by the government before a deteriorating financial situation resulted in the abandonment of the HST programme.

HST DMU set No. 253 011 (without the number painted on coupling cover) in BR InterCity 125 'rail blue' and 'rail grey' livery, with power car (later loco Class 43) No. W43022 leading, passing Yatton on a Paddington–Weston-super-Mare, October 1982. With such a large expanse of yellow predominating on the power cars, HSTs were unkindly nicknamed 'Flying Bananas' in this era.

HST DMU set No. 253 019, with power car No. W43039 (later named *The Royal Dragoon Guards*) leading, nearing Yatton on a Weston-super-Mare–Paddington service, October 1982.

With its trailers in shadow while the cab front of power car No. W43010 catches the last of the disappearing sunlight, HST set No. 253 005 departs Nailsea & Backwell station on a Paddington–Weston-super-Mare service, October 1983. The power car was formerly *TSW Today*.

Class 253 HST DMUs side by side at Paddington: on the left is set No. 253 038 in BR InterCity 125 'rail blue' and 'rail grey' livery, with leading power car No. W43017 (later named *HTV West*) having its set number in black on the coupling hatch; on the right is set No. 253 028 with leading power car No. W43125 (later named *Merchant Venturer*) in BR InterCity Executive livery with set number in white beneath the driving cab window. This was the usual position for set numbers (when used) on power cars in the InterCity Executive colours, though they fell out of use soon after the introduction of this livery. Picture taken in March 1984.

Class 253 HST set No. 253 021 in BR InterCity 125 'rail blue' and 'rail grey' livery, with leading power car No. 43 042 with no set number nearing Paddington, August 1984.

Leaving St Pancras on a Midland Main Line service is Class 253 HST set No. 253 039, with no set number, and power car number No. 43 147 in white on the guard's compartment, August 1984. The regional prefix to the number, which would have been 'M' in this instance, is no longer displayed. The loco was later named *Red Cross*, which was subsequently amended to *The Red Cross*. Set numbers were being painted out on the power cars at this time since power cars of sets they were allocated to were often replaced by power cars from other sets for operational reasons, with the result that set numbers displayed on one (or even both) power cars of a set did not match the number of the set they were actually hauling! Eventually, the confusion was cleared by reclassifying HST's from diesel multiple units (consisting of power cars and trailers) to locomotives and coaches. It had originally been intended to build ten sets dedicated to the St Pancras–Nottingham/Sheffield route, but in the event these were never ordered due to lack of funds and sets were scrounged from the Western, ECML and Cross Country routes to provide the necessary services.

Class 253 HST set No. 253 044 in BR InterCity 125 'rail blue' and 'rail grey' livery, with no set number displayed, but power car number No. 43 169 (later named *The National Trust*) in white on the cab side, at speed near Stonehouse on a North East–South West Cross Country service, May 1985. No. 43 149 was later named *BBC Wales Today*. Originally, no less than thirty-six HST sets were thought necessary to provide a comprehensive service for the various Cross Country routes, but in the event only the first tranche of eighteen sets were approved (Nos 253 041–058) and the second tranche was never authorised because of financial cut backs. Interestingly, the surprisingly large number of thirty sets had been envisaged for Trans Pennine, Edinburgh–Glasgow and other routes, but these plans were similarly abandoned.

A very grubby, anonymous Class 253 HST, with no set number on the power car and with (indecipherable) power car number in white on the side of the cab, speeding through Severn Tunnel Junction on a Paddington-bound service, August 1985. One thing to be said about rail privatisation is that these days an HST set would never be seen in such a dirty condition, though improved washing techniques have undoubtedly helped.

An unidentified Class 253 HST set approaching Parson Street station, Bristol, on a Paddington–Weston-super-Mare service, July 1985. Interestingly, the first class trailer is flanked by standard class trailers, an arrangement that would not be seen today.

In 'rail blue' and 'rail grey' livery, Class 254 HST set No. 254 034, with black set number on coupling hatch and white power car number No. 43 136 (without 'E' prefix) on the cab side at King's Cross, September 1986. The standard formation of HSTs when introduced on the ER/ScR for the East Coast Main Line service was two power cars plus eight trailers. Set numbers were being phased out by 1986, so it was fairly unusual to still see an HST with set numbers displayed. Many power cars that remained in blue and grey livery at this time had no number on the front at all, but some had the power car number in black either on the coupling cover or, less often, on the yellow strip between the grill and windscreen.

An unidentified HST in BR InterCity 125 livery with no set number but white locomotive number on the side of the driver's cab, passing Haresfield on a Cross Country service, July 1986.

HSTS UNDER BR SECTORISATION – THE INTERCITY EXECUTIVE AND SWALLOW ERA

Class 253 HST set No. 253 030 with the trailers in BR InterCity 125 'rail blue' and 'rail grey' livery, but with the leading power car (later loco Class 43) No. 43 129 in BR InterCity 125 executive livery, passing Stonehouse (Midland) on a Cross Country service, May 1985. The set number is in white below cab windscreen.

At the time (but not for much longer) still classed as a DMU, HST set No. 253 028 (but with no set number on cab front) in BR InterCity 125 Executive two-tone grey and yellow livery, with power car – later loco Class 43 – No. 43 125 *Merchant Venturer* leading, passing Parson Street, Bristol on a Paddington–Weston-super-Mare service, July 1985.

An HST with No. 43 184 in BR InterCity Executive livery and the power car/locomotive number in black on the guard's compartment and still with the set number No. 253 051 in white below the windscreen; however, it is likely that the set is a different one to that indicated on the power car since it is still in BR InterCity 125 'rail blue' and 'rail grey' livery. The HST is seen passing Haresfield on a Cross Country service, July 1986.

Class 43 No. 43 059 in BR InterCity 125 Executive livery at Leeds on a King's Cross service, July 1988. The locomotive number is both in large white numerals on the cab front below the windscreen and small black numerals on the guard's compartment. To the right is a Metro-Cammell Class 101 DMU on a service to Harrogate. When locomotives in the BR InterCity 125 Executive livery showed the loco number on the front, white numerals (of various sizes) in the grey area below the windscreen was the normal means of display.

Leaving Bristol Temple Meads on a Paddington–Weston-super-Mare service is an HST headed by Class 43 No. 43 010 (later named *TSW Today*) in BR InterCity 125 Executive two-tone grey and yellow livery, May 1989. There is no locomotive number on the front of the cab, only small numerals in black on the rear sides. The dark grey band is falcon grey, the very pale beige of the lower rear flank of the locomotive (where the guard's compartment was originally positioned) is 'light grey' and the stripes between the two colours are 'silver white' and 'rail red' (the same as on the coaches). The yellow panel is of the standard 'spectrum yellow' warning shade and has a 'silver white' stripe above.

About to depart Bristol Temple Meads on a Paddington service, Class 43 No. 43 033 (later named *Driver Brian Cooper 15 June 1947–5 October 1999*) is in BR InterCity 125 Executive two-tone grey and yellow livery, May 1989.

Class 43 No. 43 192 *City of Truro* (note small nameplate) in BR InterCity Swallow livery – in my opinion the most attractive livery ever carried by HSTs – at Bristol Temple Meads, May 1991. The power car number is in black on the guard's compartment. Apart from the italicised InterCity wording and swallow emblem, the main difference between this livery and the InterCity 125 Executive scheme is that the long warning-yellow lower panel on the front three quarters of the locomotive's flank was replaced by what appears at first sight to be the pale beige 'light grey' of the rear quarter flank, as well as the 'rail red' and 'silver white' stripes. In fact, while the rear flank remained 'light grey', the front flank is 'silver white' – the same as the stripe above. In some lights, and especially if the loco was not clean, there seemed to be no difference in the appearance of the 'light grey' and 'silver white', but this photograph clearly shows the difference in the colours and the beige tint of the 'light grey'. Note that the loco number is in small black figures on the rear 'light grey' section, but on future repaints in the Swallow livery the number was in much larger black figures placed on the front flanks below the cab windows.

Class 43 No. 43 156 (later named *Rio Champion* and then *Dartington International Summer School*) in BR InterCity 125 Executive livery arriving at Bristol Temple Meads on a Weston-super-Mare–Paddington service, May 1991. The power car number is in black next to the driver's cab door.

BR Class 43 No. 43 137 (later *Newton Abbot 150*) in BR InterCity Swallow livery at Bristol Temple Meads, May 1991. The power car number is large and in black and placed in front of the driver's cab door, which became the standard position when it was decided that the smaller numbers at the rear of the loco flanks was not sufficiently noticeable.

An unidentified HST in BR InterCity Swallow livery winding its way through the South Devon countryside and passing Totnes on a service from Paddington to Penzance, May 1992.

An HST in BR InterCity Swallow livery headed by Class 43 No. 43 165, later named *Prince Michael of Kent*, pulls into Bristol Temple Meads on a service from Paddington on a beautiful, cloudless summer day in June 1992.

Class 43 No. 43 037 (later named *Penydarren Rail Bicentenary 1804–2004*) in BR InterCity Swallow livery (with large, black power car number ahead of the driver's cab door) at Bristol Temple Meads having arrived on a service from Paddington, June 1992. In the centre road is English Electric Class 37 1,750 hp Co-Co No.37 162 (originally No. D6862) in BR railfreight grey livery with a fuel train for Bristol St Philip's Marsh MPD.

An HST with Class 43 No. 43 164 at the head passing Bedminster, Bristol, on a Paddington–Weston-super-Mare service, April 1995.

BR Class 43 No. 43 188 *City of Plymouth* in BR InterCity Swallow livery at Bristol Temple Meads on a Weston-super-Mare–Paddington service, April 1995. The difference in the 'light grey' beige rear lower flank of the locomotive and the 'silver white' of the front lower flank is readily apparent.

Class 43 No. 43 102 (formerly named *City of Wakefield*, later *HST Silver Jubilee*) in BR InterCity Swallow livery passing a rather forlorn Bedminster station, Bristol, on a rather murky day on a Weston-super-Mare–Paddington service, April 1995. The grubby state of the locomotive illustrates that, when dirty, the difference in the 'light grey' beige rear lower flank of the locomotive and the 'silver white' of the front lower flank is not easily discerned.

HSTS DURING PRIVATISATION –
GREAT WESTERN MAIN LINE

Before the First Group had taken over the franchise, Class 43 No. 43 137 *Newton Abbot 150* of Great Western in their olive green, ivory and gold livery eases out of Bristol Temple Meads on a Weston-super-Mare–Paddington service in April 1998. The green was supposedly based on the GWR's 'middle chrome green' (commonly called 'Brunswick green') but it didn't seem that this was the case to me – a Class 43 in this livery bore no comparison to a shining 'King' in 'Brunswick green'.

Descending Horfield Bank and heading towards Bristol Temple Meads on a June day in 2000 is a First Great Western HST with, at the head, Class 43 No. 43 021 (later named *David Austin – Cartoonist*). The set is in FGW's original olive green, ivory and gold livery, a modification of the livery of the previous franchisee Great Western and, in my opinion, a significant improvement. The livery was sometimes nicknamed the 'fag packet' livery!

Valenta-engined Class 43 No. 43 003 *Isambard Kingdom Brunel* in First Great Western 'swish' livery at Paddington, 11 October 2007. The trailers are in the later FGW 'dynamic lines' livery.

When First Great Western's Class 43s were refurbished, they were re-engined with German MTU engines and two, instead of three, marker lights without a glazed covering. They effectively became Class 43/0. The photograph shows refurbished No. 43 169 *The National Trust* in FGW's new livery of plain indigo blue heading an HST at Bristol Parkway on a Swansea–Paddington service, October 2007. The first repaints of FGW's refurbished Class 43/0s included the 'dynamic lines' motif of the coaches, which at least added interest to the locos' livery, but for some reason this was abandoned for subsequent repaints, much to the detriment of the aesthetics of the locos.

A First Great Western HST creeps into Bristol Parkway backlit by the early morning sun on a Swansea–Paddington service in October 2007. At the head is MTU-engined Class 43/0 No. 43 092 *Institution of Mechanical Engineers 150th Anniversary 1847–1997* (formerly named *Highland Chieftain*) in FGW plain blue livery.

Just avoiding the shadow of the canopy at Reading station, an HST of First Great Western headed by Class 43 No. 43 143 *Stroud 700* (with Valenta engine and original marker light arrangement) in 'swish' livery, catches the early morning autumn sun at Reading on a Paddington service, November 2007. The trailers are in the newer 'dynamic lines' livery.

An HST of First Great Western in their 'dynamic lines' livery headed by Class 43/0 No. 43 186 (previously named *Sir Francis Drake*) in plain blue, with MTU engine and twin light clusters, approaching Reading on a Paddington service, 17 December 2007. Note that the buffet car is still in the older blue and gold livery.

Climbing past Narroways Hill Junction, on a diverted Bristol Temple Meads–Paddington service, is a First Great Western HST with Class 43/0 No. 43 031 at the head, 30 January 2008.

First Great Western Class 43/0 No. 43 002 (formerly *Techniquest* and originally *Top of the Pops*) in plain blue livery at the rear of an HST just arrived at Cardiff from Paddington, 30 January 2008.

The early morning winter sun reflects off the flank of Class 43/0 No. 43 145 of First Great Western which is about to depart Paddington, 17 February 2008. The trailers are former Midland Main Line coaches (de-branded) on hire to FGW at a time when they were short of stock because of FGW's refurbishment programme of its own stock.

Class 43/0 No. 43 140 of First Great Western in their plain blue livery entering Cardiff Central on a service from Paddington, 26 March 2008. The de-branded Midland Mainline trailers are on hire to FGW due to a shortage of stock brought about by FGW refurbishing its own stock.

A First Great Western HST headed by Class 43/0 No. 43 146 speeding through Hayes & Harlington as it heads towards Paddington, 7 April 2008.

An HST of First Great Western in their 'dynamic lines' livery with, at the rear, Class 43/0 No. 43 168 at Newport High Street on a Swansea High Street–Paddington service, 26 March 2008.

A First Great Western HST in their 'dynamic lines' livery with Class 43/0 No. 43 193 at the tail passing Abbey Wood, Bristol, on a diverted Paddington–Bristol Temple Meads service, May 2008. No.43 193 was originally named *Yorkshire Post*, then *Plymouth Spirit of Discovery* and finally *Rio Triumph*.

Class 43/0 No. 43 177 brings up the rear of a First Great Western HST leaving Westbury on a Plymouth–Paddington service, 8 May 2008.

Class 43/0 No. 43 020 (formerly named *John Grooms*) of First Great Western in plain blue livery with the coaches in 'dynamic lines' blue livery on the Westbury cut-off with a Paddington-bound service, 8 May 2008.

An unidentified HST approaching Westbury station from Fairwood Junction with a Paddington service, 8 May 2008.

Class 43/0 No. 43 003 *Isambard Kingdom Brunel* of First Great Western negotiating Fairwood Junction and heading for Westbury station with a Plymouth–Paddington service, 8 May 2008.

A First Great Western HST with Class 43/0 No. 43 130 (ex-*Sulis Minerva*) in the rear having just passed Fairwood Junction on its way to Westbury station with a Plymouth–Paddington service, 8 May 2008.

A First Great Western HST with Class 43/0 No. 43 063 (originally named *Maiden Voyager*, then *Rio Challenger*) bringing up the rear, leaving Westbury with a Paddington–Plymouth service, 8 May 2008.

An HST of First Great Western with Class 43/0 No. 43 137 (originally named *Newton Abbott 150*) bringing up the rear, at Westbury with a Paddington–Plymouth service, 8 May 2008.

Class 43/0 No. 43 189 (previously named *Railway Heritage Trust*) of First Great Western at the head of an HST arriving at Westbury with a Paddington–Plymouth service, 8 May 2008.

Class 43/0 No. 43 188 (ex-*City of Plymouth*) of First Great Western at Bristol Temple Meads on a Paddington service, 5 July 2008.

Stopping at Plymouth on a Paddington–Penzance service on 18 July 2008 is a First Great Western HST headed by blue Class 43/0 No. 43 087 (previously *Rio Invader* and later *11 Explosive Ordnance Disposal Regiment Royal Logistic Corps*), which had been the Hornby-liveried loco.

Bringing up the rear of a First Great Western HST newly arrived from Paddington, Class 43/0 No. 43 140 basks in the summer gloom of a drizzly July day at Plymouth 18 July 2008. Plymouth (originally Plymouth North Road) was completed in 1877 as a joint station of the Great Western Railway and the London & South Western Railway. Today, it is, in its way, a typical example of 1960s station architecture on BR, being completed in 1962 – although the rebuilding was first begun in 1938. The Second World War intervened.

First Great Western Class 43/0 No. 43 093 (formerly *Lady in Red* and originally *York Festival '88*) is at the rear of an HST departing Plymouth for Penzance, 18 July 2008.

Class 43/0 No. 43 124 (formerly *BBC Points West*) of First Great Western speeding through Keynsham with a Paddington–Bristol Temple Meads service, 26 July 2008.

First Great Western Class 43/0 No. 43 023 (ex-*County of Cornwall*) approaching Keynsham with a Bristol Temple Meads–Paddington service, 26 July 2008.

Class 43/0 No. 43 156 *Dartington International Summer School* (ex-*Rio Champion*) of First Great Western at Bristol Temple Meads on a Paddington bound service, 30 September 2008. Note the eyesore of the derelict Royal Mail sorting office which has become something of a welcoming landmark for travellers arriving at Bristol! The equally attractive overbridge – which seriously defaces the station – is a disused link between the sorting office and the platforms which transported mail in the heady days when mail was carried by trains to and from Bristol. With the impending refurbishment of Temple Meads, it is to be hoped that this bridge is removed.

An HST of First Great Western headed by Class 43/0 No. 43 190 passing Patchway on a Paddington–South Wales service, March 2009.

In glorious Berkshire, Class 43/0 No. 43 027 *Glorious Devon* (first named *Westminster Abbey*) of First Great Western glints beneath the high noon summer sun on a hot day at Reading, 23 June 2009.

An HST headed by Class 43/0 No. 43 187 of First Great Western in Bourton Cutting near Flax Bourton on a Paddington–Weston-super-Mare service, 2 August 2009. It was at this location on 13 October 1876 that the Up *Flying Dutchman*, hauled by a broad gauge locomotive, derailed, climbed the bank and fell back onto the line, turning completely over lengthwise, with the coaches behind then ploughing into the engine. Newspapers at the time reported that the fireman was 'almost disembowelled' and had 'the back of his head cut off'. The driver had both an arm and a leg cut off. Both, unsurprisingly, died. Amazingly, though several passengers were injured, none lost their life.

First Great Western Class 43/0 No. 43 033 *Driver Brian Cooper 15 June 1947–5 October 1999* in Bourton Cutting near Flax Bourton on a Paddington–Weston-super-Mare service, 2 August 2009.

First Great Western Class 43/0 No. 43 087 (previously named *Rio Invader* and subsequently named *11 Explosive Ordnance Disposal Regiment Royal Logistic Corps*) accelerating past Flax Bourton on a Weston-super-Mare–Paddington service, 2 August 2009.

Class 43/0 No. 43 150 (ex-*Bristol Evening Post*) brings up the rear of a First Great Western HST passing Bedminster, Bristol, on a Weston-super-Mare–Paddington service, 9 August 2009.

First Great Western Class 43/0 No. 43 004 *First of the Future/First Ar Gyfer y Dyfadol* (formerly *Borough of Swindon* and originally *Swan Hunter*) passing Ram Hill, Coalpit Heath, on a Paddington–Swansea service, 22 August 2009.

Class 43/0 No. 43 186 *Sir Francis Drake* of First Great Western glides past Ram Hill, Coalpit Heath, on a Swansea–Paddington service, 22 August 2009.

A First Great Western HST headed by Class 43/0 No. 43 088 (first named *XIII Commonwealth Games Scotland 1986*, then *Rio Campaigner*) passing Coalpit Heath on a South Wales–Paddington service, 22 August 2009.

Class 43/0 No. 43 037 (formerly *Penydarren Rail Bicentenary 1804–2004*) of First Great Western passing Ram Hill, Coalpit Heath, on a South Wales-bound service, 22 August 2009.

A First Great Western HST at Didcot Parkway on a Swansea–Paddington service, 16 September 2009. Class 43/0 No. 43 153 (originally named *University of Durham*, then *The English Riviera Torquay, Paignton, Brixham*) is at the tail.

First Great Western HST headed by Class 43/0 No. 43 032 (formerly *The Royal Regiment of Wales*) at speed near Yatton on a Paddington–Weston-super-Mare service, 23 August 2009.

Basking in late afternoon winter sunshine, an HST of First Great Western headed by Class 43/0 No. 43 193 arriving at Bristol Temple Meads with a service from Paddington, 4 January 2010. No. 43 193 was first named *Yorkshire Post*, then *Plymouth City of Discovery*, and finally *Rio Triumph*.

Class 43/0 No. 43 181 of First Great Western (formerly *Devonport Royal Dockyard 1693–1993*) at Bristol Temple Meads on a Paddington service, 8 April 2010.

Class 43/0 No. 43 056 (formerly *University of Bradford*) of First Great Western coming into Bristol Temple Meads on a service from Paddington, 8 April 2010.

A First Great Western HST with Class 43/0 No. 43 165 *Prince Michael of Kent* in the lead mounting Pilning Bank and approaching Patchway station on a South Wales–Paddington service, May 2010.

Class 43/0 No. 43 145 of First Great Western nearing Keynsham on a Paddington–Bristol Temple Meads service, 29 August 2010.

Class 43/0 No. 43 138 of First Great Western between Keynsham and Bath on a Paddington–Bristol Temple Meads service, 19 March 2011.

First Great Western Class 43/0 No. 43 087 *11 Explosive Ordnance Disposal Regiment Royal Logistic Corps* (ex-*Rio Invader*) between Keynsham and Bath on a Bristol Temple Meads–Paddington service, 19 March 2011.

With Parson Street station, Bristol, beyond the bridge in the background, Class 43/0 No. 43 136 of First Great Western heads towards Bristol Temple Meads on a Weston-super-Mare–Paddington service, 6 May 2011.

Class 43/0 No. 43 192 (ex-*City of Truro*) of First Great Western passing through Sydney Gardens and approaching Bath on a Paddington–Weston-super-Mare service, 14 September 2011.

With the maize growing tall in the adjoining field, First Great Western Class 43/0 No. 43 094 between Keynsham and Bath on a Bristol Temple Meads–Paddington service, 28 September 2011.

Class 43/0 No. 43 193 (first named *Yorkshire Post*, then *Plymouth City of Discovery*, and lastly *Rio Triumph*) of First Great Western on a Weston-super-Mare–Paddington service approaching Bath, 28 September 2011.

With the long shadows of a setting sun falling across the platforms, Class 43/0 No. 43 017 (ex-*BBC Points West*) at the rear of a First Great Western HST passes through Patchway on a Swansea–Paddington service, 3 January 2012.

Class 43/0 No. 43 189 *Railway Heritage Trust* heads a First Great Western HST on a Bristol Temple Meads–Paddington service passing another HST on a Paddington–Bristol Temple Meads service at Barton Hill, 22 July 2012. In the distance is a Cross Country Class 221 Super Voyager heading northwards.

With the evening sun highlighting the locomotive, a First Great Western HST headed by Class 43/0 (MTU) No. 43 130 (ex-*Sulis Minerva*) speeding past the site of the long-closed Flax Bourton station on a Paddington–Weston-super-Mare service, 22 July 2012.

Class 43/0 No. 43 134 (ex-*County of Somerset*) of First Great Western on a Paddington–Hereford service at Moreton-in-Marsh, 2 July 2012.

Platforms 12/13 and 14/15 of Bristol Temple Meads are the platforms where First Great Western's Paddington services start from or terminate. Here, on 11 April 2014, are two HSTs headed by Class 43/0 Nos 43 088 (ex-*Rio Campaigner*, ex-*XIII Commonwealth Games Scotland 1986*) and 43 142 *Reading Panel Signal Box 1965–2010* (ex-*St Mary's Hospital Paddington*). The disused Royal Mail building in the background is in an even more derelict condition than it is in the view of it on page 43.

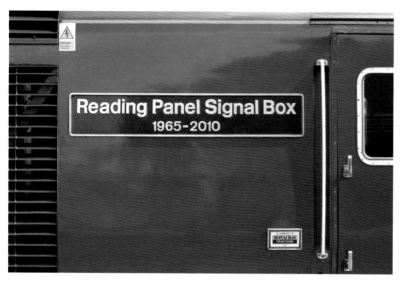

Nameplate of BR Class 43/0 No. 43 142 *Reading Panel Signal Box 1965–2010* (previously *St Mary's Hospital Paddington*) at Bristol Temple Meads, 11 April 2014. Modern First Great Western locomotive names do not seem to be quite so evocatively named as were GWR steam locomotives – *Knight of the Golden Fleece*, *Lady of Quality*, *The Wolf* and *Thunderbolt* spring to mind as fine examples from the days of Dean & Churchward, although it has to be said that the endless lists of stately homes in the days of Collett were not so evocative. Note, bottom right, the Brush Works Plate indicating the locomotive's refurbishment with a MTU engine in 2007.

43/0 No. 43 070 (formerly *The Corps of Royal and Electrical Engineers*, originally *Rio Pathfinder*) of First Great Western between Nailsea & Backwell and Yatton on a Paddington–Weston-super-Mare express, 18 April 2014.

On a beautiful sunny morning, Class 43/0 No. 43 091 (formerly *Edinburgh Military Tattoo*) of First Great Western speeds past a level crossing between Nailsea & Backwell and Yatton on a Paddington–Weston-super-Mare express, 18 April 2014.

Pictured between Yatton and Nailsea & Backwell is a First Great Western HST with Class 43/0 No. 43 069 (formerly named *Rio Enterprise*) at the rear on a Weston-super-Mare–Paddington service, 21 May 2014.

First Great Western's Class 43/0 No. 43 162 (ex-*Borough of Stevenage*) arrives at Didcot Parkway with a Swansea–Paddington service, 25 May 2014.

HSTS DURING PRIVATISATION –
MIDLAND MAIN LINE

Midland Mainline Class 43 No. 43 070, as built with Ventura engine and the original glazed coverings to light clusters, in MMl's later blue and two-tone grey livery at Stockport on a St Pancras–Manchester (Piccadilly) service, at a time when extensive engineering work on the WCML necessitated its part-closure and the replacement of electric Euston–Manchester services by HSTs from St Pancras, September 2004. This replacement service was named *Project Rio* and the locomotives used on this service had an 'R' placed above the loco number, as on No. 43 070. The locos also carried a name prefixed by *Rio*, although No. 43 070 is yet to be so named. (It was to be named *Rio Pathfinder* and when that was subsequently removed was then named *The Corps of Royal & Electrical Engineers*.)

Class 43/0, with a Paxman 12VP185 engine, No. 43 043 (ex-*Leicestershire County Cricket Club*) of East Midlands Trains in de-branded Midland Mainline livery arriving at Derby with a service from St Pancras, 13 April 2008.

An East Midlands HST in de-branded Midland Mainline livery departing Sheffield and heading for St Pancras, 30 April 2008. The rear of the train is brought up by a smoking, Paxman 12VP185-engined, Class 43/0 No. 43 043, originally named *Leicestershire County Cricket Club*.

Class 43/0 No. 43 050, with a Paxman 12VP185 engine and triple light clusters, of East Midlands Trains in unbranded Midland Mainline livery at Loughborough on a St Pancras service, 15 July 2008. Behind is Brush's Falcon Works, which fitted the German MTU engines to refurbished Class 43s.

Paxman 12VP185-engined Class 43/0 No. 43 076 (originally *BBC East Midlands Trains*, then *The Master Cutler 1947–1997*) of East Midlands Trains in unbranded Midland Mainline livery entering Derby on a service from St Pancras, 15 July 2008.

Class 43/0 No. 43 050 of East Midlands Trains in unbranded Midland Mainline livery at the head of an HST at Nottingham on a St Pancras service, 15 July 2008.

Following the rebuilding of St Pancras to cater for Eurostar, the Midland trains have now been banished to new, characterless platforms at the north-western corner of the station. Gone are the days of getting a picturesque shot of an HST, let alone a *Peak* or, dare I say, a *Jubilee*, under Barlow's impressive train shed. Pictured at St Pancras on a Nottingham service is Class 43/0 No. 43 052 (ex-*City of Peterborough*) of East Midlands Trains, 21 January 2010. The locomotive has been refurbished with a 2,100 hp Paxman 12VP185 engine and triple light clusters; it is in EMTs, then new livery, which was the standard livery of the parent company Stagecoach. The livery is rather reminiscent in style to FGW's blue 'swish' livery, as the multi-coloured 'swoops' on the front flank of the Class 43 seem to make the locomotive appear as if it's going fast even when stationary. In my view, a very attractive livery.

An East Midlands Trains HST in their full livery headed by 12VP185-engined Class 43/0 No. 43 044 (previously named *Borough of Kettering*) passing Chesterfield southbound for St Pancras, 8 February 2014.

Class 43/0 No. 43 044 (previously named *Borough of Kettering*) of East Midlands Trains departing St Albans (City) on a Leeds–St Pancras service, as a Class 319 EMU passes on a Moorgate–Bedford service, 15 April 2014.

Class 43/0 No. 43 052 (previously named *City of Peterborough*) of East Midlands Trains in the standard Stagecoach livery approaching St Albans on a St Pancras–Nottingham service, 15 April 2014.

Passing the hamlet of Sandridgebury, to the north of St Albans, on a St Pancras–Nottingham service is an East Midlands Trains HST headed by Class 43/0 No. 43 044 (previously named *Borough of Kettering*), 15 April 2014.

East Midlands Trains Class 43/0 No. 43 046 (originally named *Royal Philharmonic*) speeds past Sandridgebury on a St Pancras–Nottingham service, 15 April 2014.

Class 43/0 No. 43 059 of East Midlands Trains at Sandridgebury on a St Pancras–Nottingham service, 15 April 2014.

Sweeping around the curves at East Hyde is Class 43/0 No. 43 044 (previously named *Borough of Kettering*) on a Nottingham–St Pancras service, 15 April 2014.

Speeding past East Hyde is Class 43/0 No. 43 052 (previously named *City of Peterborough*) of East Midlands Trains on a St Pancras–Nottingham service, 15 April 2014.

Passing under an impressive signal gantry at East Hyde is East Midlands Trains Class 43/0 No. 43 073 on a St Pancras–Nottingham service, 15 April 2014. The lines on the locomotive are shadows from the overhead lines.

East Midlands Trains Class 43/0 No. 43 059 on a Nottingham–St Pancras service passing East Hyde, 15 April 2014.

HSTS DURING PRIVATISATION –
EAST COAST MAIN LINE

Class 43 No. 43 039 *The Royal Dragoon Guards* of Great North Eastern Railway at Kings Cross on an Aberdeen service, 21 July 2007. No. 43 0139 still has a Paxman Valenta engine but, being a former Midland Mainline loco, has received replacement independent triple light clusters rather than the usual twin clusters with which GNER fitted its Class 43s. In my view, the GNER livery of a deep, midnight-blue base with orange-red stripe and 'retro-reflective' gold GNER lettering in Futura font was one of the more attractive liveries of first generation privatisation. Interestingly, the blue was the same colour as that used by Wagon Lits.

When the Great North Eastern Railway refurbished their Class 43s with MTU engines, the locomotives were renumbered by adding 200 to their original number and then reclassified as 43/2. Pictured here is No. 43 306 *Fountains Abbey* (formerly No. 43 106 *Songs of Praise*) in GNER deep-blue livery with an orange-red stripe and also with a Leeds–London Service advertisement pulling up to the buffers at Kings Cross, rather inappropriately on a service from Aberdeen, not Leeds, on 21 July 2007. Several GNER Class 43s carried advertisements at this time.

A National Express East Coast HST creeps out of Gas Works Tunnel into Kings Cross, 22 February 2008. The trailers are repainted in full NXEC livery but the locomotive is in Interim NXEC livery, with the GNER red-orange stripe replaced by a NXEC white stripe. The Class 43/2, which has been refurbished with a MTU engine and twin light clusters, is No. 43 208 (ex-43 008 *City of Aberdeen*).

Valenta-engined, buffer-fitted Class 43 No. 43 080 of Grand Central in their surprisingly attractive all-black livery arriving at Kings Cross on a service from Sunderland, 22 February 2008.

Class 43 No. 43 077 (ex-*County of Nottingham*) of National Express East Coach in interim NXEC livery. No. 43 077 was a former Midland Mainline loco refitted with triple light clusters but retaining a Valenta engine; however, GNER soon replaced that with a MTU engine. The locomotive is seen at Kings Cross on an Aberdeen service, 22 February 2008.

Black, Valenta-engined Class 43 No. 43 080, fitted with buffers, of Grand Central entering Kings Cross on a service from Sunderland, 22 February 2008.

MTU-engined Class 43/2 No. 43 367 (ex-No.43 167) *Deltic 50 1955–2005* in full National Express East Coast livery entering Kings Cross, 30 June 2008. The coaches are in interim NXEC livery.

Refurbished Class 43/2 No. 43 309 (formerly No. 43 109 *Scone Palace*, ex-*Yorkshire Evening Post*) of National Express East Coast in their then new livery at Kings Cross on an Aberdeen service, 30 June 2008.

Valenta-engined, buffer-fitted, Class 43 No. 43 084 of Grand Central in their attractive all-black livery at Kings Cross on a Sunderland service, 30 June 2008.

Valenta-engined, buffer-fitted, Class 43 No. 43 123 of Grand Central in black livery arriving at Kings Cross on a Sunderland service, 30 June 2008. Note damaged warning horn grill.

When Grand Central were formed, they had to scour the dregs of stored Class 43s, some of which were little more than wrecks, and had them refurbished by DML of Devonport Dockyard, surprisingly still with Paxman 12RP200 Valenta engines, albeit renovated, and with new twin, recessed marker light clusters. Insufficient off-lease HST trailers were available so a number of locomotive-hauled coaches were converted to HST trailers.

Seen under Lewis Cubitt's roof, a Grand Central HST has just arrived at Kings Cross on a service from Sunderland, with Valenta-engined, buffer-fitted Class 43 No. 43 084 (ex-*County of Derbyshire*) at the rear, 30 June 2008.

Class 43/2 No. 43 318 (formerly No. 43 118 *City of Kingston-upon-Hull* and originally *Charles Wesley*) in interim National Express East Coast livery zooming through Doncaster on an Aberdeen–Kings Cross service, 2 September 2008.

With a railway enthusiast taking a photograph, a Grand Central HST, with Class 43 No.43 080 at its head, passes Doncaster on a King's Cross–Sunderland service, 2 September 2008. Buffer-fitted No. 43 080 still has a Valenta engine but has been refitted with twin light clusters.

Valenta-engined Class 43 No. 43 065 (formerly named *City of Edinburgh*), with buffers, of Grand Central speeding through Doncaster on a Kings Cross–Sunderland service, 2 September 2008.

A National Express East Coast HST in their interim livery with Class 43/2 No. 43 108 (first named *BBC Television Railwatch*, then *Old Course St Andrews*) speeding through Doncaster on a Kings Cross–Aberdeen service, 2 September 2008.

Class 43 (Valenta engined) No. 43 080 (with buffers) of Grand Central winds its way around the rear of Newcastle station on its way to Heaton depot from Sunderland (where the ex-Kings Cross service it would have formed would have been terminated), 2 September 2008.

Buffer fitted, Valenta-engined Class 43 No. 43 065 (formerly *City of Edinburgh*) of Grand Central passing Newcastle station on its way to Heaton depot, 2 September 2008.

Class 43/2 No. 43 206 (ex-No.43 006 *Kingdom of Fife*) of National Express East Coast in their interim livery at the head of a consist with the trailers mostly in full NXEC livery but one in interim livery, leaving Newcastle on a Kings Cross–Aberdeen service, 2 September 2008.

Class 43/2 No. 43 305 (previously No.43 105 *City of Inverness* originally *Hartlepool*) of National Express East Coast in their full livery at the head of a consist of NXEC interim liveried trailers, arriving at Newcastle on an Aberdeen–Kings Cross service, 2 September 2008.

An HST of National Express East Coast uniformly in their full livery headed by Class 43/2 No. 43 312 (ex-No. 43 112 *Doncaster*) at Newcastle on a Kings Cross–Aberdeen service, 2 September 2008.

An HST of East Coast in their branded National Express East Coast livery at Kings Cross, 17 November 2009. At the head is Class 43/2 No. 43 314 (formerly No. 43 114 *National Garden Festival Gateshead*; originally *East Riding of Yorkshire*).

An HST headed by Class 43/2 No. 43 296 (previously No. 43 096 *Stirling Castle*; originally *The Queen's Own Hussars*) of East Coast in rebranded former National Express East Coast livery leaving York on a King's Cross–Aberdeen service 15 September 2010.

Nameplate of East Coast Class 43 No. 43 367 (ex-No.43 167) *Deltic 50 1955–2005* at Kings Cross, June 2011.

East Coast Class 43/2 No. 43 272 (ex-No. 43 072 *Derby Etches Park*) in their full livery at Kings Cross, 30 August 2012. This loco is ex-Midland Mainline/East Midlands Trains, hence the triple light clusters rather than twin. Note trapezoidal black painted area around the driver's side window, which seems to be the usual shape on ex-MMl locos. The trailers are still in National Express East Coast livery, albeit rebranded East Coast.

Class 43/2 No. 43 309 (formerly No. 43 109 *Scone Palace*; originally *Yorkshire Evening Post*) of East Coast in their full livery at Kings Cross, 30 August 2012. Note that the black area surrounding the driver's side window is in the form of a triangle, which is the standard shape for twin light clustered East Coast Class 43s.

East Coast Class 43/2 No. 43 314 (previously No.43 114 *East Riding of Yorkshire* and originally named *National Garden Festival Gateshead 1990*) is seen between Hatfield and Welham Green on a Skipton–Kings Cross service, as another HST passes in the opposite direction heading for Newark North Gate, 16 April 2014. While the locomotive is in full East Coast livery, the coaches are still in National Express East Coast livery, albeit rebranded East Coast.

Class 43/2 No. 43 274 (ex-No. 43 074) of East Coast between Hatfield and Welham Green on a Harrogate–Kings Cross service, 16 April 2014. The triple marker light clusters indicate this is a former Midland Mainline loco.

A backlit Class 43/2 No. 43 319 (ex-No. 43 119 *Harrogate Spa*) of East Coast at speed between Hatfield and Welham Green on a Harrogate–Kings Cross service, 16 April 2014.

Class 43/2 No. 43 468 (formerly No. 43 068 *The Red Nose*; originally *The Red Arrows*) of Grand Central in their revised black livery with an orange stripe approaching Welham Green at speed on a Sunderland–Kings Cross service, 16 April 2014. After reliability problems with the renovated Valenta-engined Class 43s as originally refurbished, Grand Central decided to re-engine all their fleet with more up-to-date, tried and tested, MTU engines.

East Coast Class 43/2 No. 43 319 (ex-No. 43 119 *Harrogate Spa*) passes through Welham Green station on a Kings Cross–Leeds service, 16 April 2014.

At Welham Green station is Class 43/2 No. 43 274 (ex-No. 43 074) of East Coast on a Kings Cross–Leeds service, 16 April 2014. This was a former Midland Mainline locomotive, hence the triple marker light clusters.

Grand Central Class 43/2 No. 43 423 *Valenta 1972–2010* (ex-No. 43 123) in black livery with orange stripe approaching Welham Green on a Kings Cross–Sunderland service, 16 April 2014.

Grand Central Class 43/2 No. 43 468 (ex-No. 43 068 *The Red Nose*; originally *The Red Arrows*) flashes through Welham Green station on a Kings Cross–Sunderland service, 16 April 2014.

Unusually, there is no number on the coupling cover of Class 43/2 No. 43 239 (previously No. 43 039 *The Royal Dragoon Guards*) of East Coast, which is seen approaching Welham Green on a Kings Cross–Inverness service, 16 April 2014.

East Coast Class 43/2 No. 43 296 (formerly No. 43 096 *Stirling Castle*, originally *The Queen's Own Hussars*) heads towards Welham Green on an Edinburgh (Waverley)–Kings Cross service, 16 April 2014.

Seen at Welham Green station on a Newark (North Gate)–Kings Cross service is East Coast Class 43/2 No. 43 238 (ex-No. 43 038 *City of Dundee*, originally *National Railway Museum The First Ten Years*), 16 April 2014.

Passing through Potters Bar station on a York–Kings Cross service is Class 43/2 No. 43 311 (ex-43 111 *Scone Palace*) of East Coast, 16 April 2014.

Class 43/2 No. 43 311 (ex-43 111 *Scone Palace*) of East Coast speeds towards Brookman's Park on on a Kings Cross–Edinburgh (Waverley) service, 16 April 2014.

East Coast Class 43/2 No. 43 295 (previously No. 43 095 *Perth*, originally *Heaton*) approaches Brookman's Park on a Kings Cross–York service, 16 April 2014.

Passing Brookman's Park is Class 43/2 No. 43 302 (formerly No. 43 102 *HST Silver Jubilee*, originally *City of Wakefield*) is on an East Coast Newark (North Gate)–Kings Cross service, 16 April 2014.

On a Sunderland–Kings Cross service, Class 43/2 No. 43 480 (previously No. 43 080 *The Red Nose*; originally *The Red Arrows*) of Grand Central in their black and orange livery speeds past Brookman's Park, 16 April 2014.

Class 43/3 No. 43 208 (ex-No. 43 008 *Lincolnshire Echo* and originally *City of Aberdeen*) of East Coast in EC-branded National Express East Coast livery, one of the last in such an interim livery at this time. It is seen speeding towards Harringay on a Harrogate–Kings Cross service, 7 May 2014.

Class 43/0 No. 43 073 of East Midlands Trains in the standard Stagecoach livery on hire to East Coach due to a shortage of HST sets. The HST is seen on a Skipton–Kings Cross service approaching Harringay at speed, overtaking a Class 313 on a local service, 7 May 2014.

Class 43/2 No. 43 300 (formerly No. 43 100 *Craigentinny*, previously *Blackpool Rock* but originally named *Craigentinny*) of East Coast nears Harringay on a Kings Cross–Leeds service, 7 May 2014.

East Coast Class 43/2 No. 43 307 (ex-No. 43 107 *Tayside*, originally *City of Derby*) leans into the steeply canted curve through Harringay on an Edinburgh (Waverley)–Kings Cross service, 7 May 2014.

East Coast Class 43/2 No. 43 311 (ex-No. 43 111 *Doncaster*, originally *Scone Palace*) is captured between two signal posts as it leans into the steeply graded curve through Harringay on an Edinburgh (Waverley)–Kings Cross service, 7 May 2014.

Class 43/2 No. 43 367 (ex-No. 43 167 *Deltic 50 1955–2005*) of East Coast approaches Hornsey on a Kings Cross–Inverness service, 7 May 2014.

East Coast Class 43/2 No. 43 308 (ex-No. 43 108 *St Andrews*, previously *Old Course St Andrews* and originally *BBC Television Railwatch*) speeds through Hornsey on a Kings Cross–Inverness service, 7 May 2014.

Class 43/2 No. 43 368 (ex-No. 43 068 *The Red Nose* and originally *The Red Arrows*) of Grand Central in their revised black livery with orange stripe climbing towards Alexandra Palace on a Kings Cross–Sunderland service, 7 May 2014.

HSTS DURING PRIVATISATION –
CROSS COUNTRY

Class 43 No. 43 184 (later No. 43 384) of Cross Country in de-branded Midland Mainline livery and yet to receive XC branding, passing Abbey Wood, Bristol, May 2008. The locomotive still has a Valenta engine and the original glazed marker light clusters.

Neatly captured between two signal posts is a Class 43 loco of Cross Country in de-branded Midland Mainline livery on a Penzance–Dundee service departing Bristol Temple Meads, 23 May 2008. The loco is No. 43 184, retaining a Valenta engine and original light clusters.

An HST headed by Class 43 No. 43 007 (later No. 43 207), still with a Valenta engine and original light clusters, of Cross Country in de-branded Midland Mainline livery on a Penzance–Dundee service under Francis Fox's 1870s train shed, Bristol Temple Meads, 23 May 2008.

An HST of National Express East Coast in their interim GNER-based livery on hire to Cross Country, arriving at Bristol Temple Meads on a Manchester Piccadilly–Newquay 'Surfers' Special', 24 May 2008. At the rear is Class 43 No. 43 108 (originally named *BBC Television Railwatch*, then *Old Course at St Andrews* and lastly *St Andrews*), still with a Valenta engine but with revised twin light clusters.

Class 43/2 No. 43 309 (ex-No. 43 109) (formerly *Scone Palace*, and originally *Yorkshire Evening Press*), with MTU engine and twin light clusters, of National Express East Coast in their full livery, although the rest of the set is in interim, GNER-based, NXEC livery. The HST is on hire to Cross Country and is drawing into Bristol Temple Meads on a Manchester Piccadilly–Newquay 'Surfers' Special', 24 May 2008.

Hired by Cross Country is a National Express East Coast HST headed by Class 43/2 No. 43 309 (formerly No. 43 109 *Scone Palace*, and originally *Yorkshire Evening Press*), refurbished with MTU engine and twin light clusters. The HST is on a Manchester Piccadilly–Newquay 'Surfers' Special' and is seen at Bristol Temple Meads, 24 May 2008.

Valenta-engined Class 43 (with twin light clusters) No. 43 108 (ex-*St Andrews*, previously *Old Course at St Andrews* and originally *BBC Television Railwatch*) of National Express East Coast in their interim livery on hire to Cross Country on a Manchester Piccadilly–Newquay 'Surfers' Special' departing Bristol Temple Meads, 24 May 2008.

An HST of National Express East Coast in their interim livery on hire to Cross Country passes through Stapleton Road, Bristol on a Paignton–Newcastle service, 24 May 2008. In the rear is Class 43/2 No. 43 206 (formerly No. 43 006 *Kingdom of Fife*), with MTU engine and twin light clusters.

An HST headed by Class 43 No. 43 007 (later No. 43 207), with a Valenta engine and original light clusters, of Cross Country in de-branded Midland Mainline livery on a Newcastle–Newquay Summer service passing Abbey Wood, Bristol, 24 May 2008.

Class 43/3 No. 43 317 (ex-No. 43 117 *Bonnie Prince Charlie*), with MTU engine and twin light clusters, of National Express East Coast in their full livery but with the coaches in NXEC's interim livery, on hire to Cross Country arriving at Bristol Temple Meads on a Manchester–Newquay service, 30 August 2008.

Class 43 No. 43 053 (formerly *Leeds United* and originally *County of Humberside*) of National Express East Coast in interim NXEC livery on hire to Cross Country at Bristol Temple Meads on a Manchester–Newquay service, 30 August 2008. The loco, formerly of Midland Main Line, is Valenta engined and has triple light clusters.

Bathing in the late afternoon autumn sunshine, Class 43/2 No. 43 301 (previously No. 43 101 *The Irish Mail* and originally *Edinburgh International Festival*) of Cross Country at Bristol Parkway, September 2008. The MTU-engined loco is in full XC livery but the coaches are in unbranded Midland Mainline livery.

An HST of Cross Country in their full livery headed by MTU-engined Class 43/2 No. 43 384 passing Abbey Wood, Filton on a Leeds–Plymouth service, 15 August 2009.

An HST of Cross Country departing Bristol Temple Meads on a southbound service, 22 August 2009; at the rear of the formation is Class 43/2 No. 43 303 (ex-No. 43 103) which was originally named *John Wesley*, and later renamed *Helston Furry Dance*.

Class 43/2 No. 43 384 (ex-No. 43 184) of Cross Country leans into the curve as it accelerates past Ram Hill, Coalpit Heath, on a south-bound service, 22 August 2009.

Class 43/2 No. 43 366 (ex-43 166) of Cross Country speeding past Ram Hill, Coalpit Heath, on a north-bound service, 22 August 2009.

Passing Ram Hill, Coalpit Heath, Class 43/2 No. 43 285 (ex-43 085 *City of Bradford*) of Cross Country is on a north-bound service, 22 August 2009.

Cross Country Class 43/2 No. 43 207 (ex-No. 43 007) mounting Horfield Bank and approaching Abbey Wood, Bristol on a north-bound service, July 2010.

An unidentified Cross Country HST approaching Parson Street Junction, Bristol, on a northbound express, 6 June 2011. On the right is the Bristol Freightliner Depot with No. 66 561 in the yard. On the far right is the Royal Portbury Dock branch.

A Cross Country HST pulls into Chesterfield on a Plymouth–Leeds service. At the rear is Class 43/2 No. 43 285 (ex-43 085 *City of Bradford*), 14 April 2012.

Class 43/2 HST No. 43 303 (ex-43 103 *Helston Furry Dance*, ex-*John Wesley*) of Cross Country leaving Chesterfield on a southbound service, 8 February 2014.

Class 43/2 No. 43 321 (ex-No. 43 121 *West Yorkshire Metropolitan County*) of Cross Country between Nailsea & Backwell and Yatton on a Leeds–Plymouth service, 18 April 2014.

Cross Country's Class 43/2 No. 43 357 (ex-No. 43 157 *HMS Penzance*, originally *Yorkshire Evening Post*) between Nailsea & Backwell and Yatton on a Leeds–Plymouth service, 18 April 2014.

Cross Country's Class 43 No. 43 207 (ex-43 007) emerges out of Wickwar Tunnel, north of Yate, on a Dundee–Plymouth service, 18 April 2014.

Class 43/2 No. 43 357 (ex-43 157 *HMS Penzance*, originally *Yorkshire Evening Post*) of Cross Country speeds through Wickwar Cutting on a Plymouth–Glasgow (Central) service, 18 April 2014.

Heading towards Wickwar Tunnel on a Plymouth–Glasgow (Central) service is Class 43/2 No. 43 321 (ex-43 121 *West Yorkshire Metropolitan County*) of Cross Country, 18 April 2014.

POWER CARS/LOCOMOTIVES – A COMPARISON

The prototype power cars/locomotives of what was originally known as the High Speed Diesel Train differed significantly from the production version. One of the former, Class 41 No. 41 001, is pictured at the Didcot Railway Centre Diesel Gala, 25 May, 2014. It is in original 'reverse Inter City rail grey and rail blue' livery. Note that the white loco number does not stand out against the pale grey background.

The prototype's cab was very different to the production version, being designed for one man operation with a central driving position. The front consisted of a driver's windscreen below which was a separate windscreen covering the marker lights. No side windows were fitted to the cab. Buffers were fitted as it was originally envisaged that, as locomotives, they might also be used to haul non-HSDT trains such as sleeper trains and even high speed freight trains! The rear compartment of the locomotive housed not simply accommodation for the conductor (i.e. guard) and luggage, but an auxiliary driving position with simplified controls and a small windscreen to one side of the gangway connection to the coaches. In addition, on each side of the guard's accommodation was an access door with window and another window to the rear of that. When the prototype HSDT was later classified as a DMU (set No. 252 001), the two locomotives were classified as DMB's – Driving Motor Brakes. Locomotive No. 41 001 became power car No. W43000.

No. 41 001 has now been completely restored by the 125 Group as part of 'Project Miller', named after the British Rail Chief Engineer of Traction and Rolling Stock, Terry Miller, who led the design team for the HST. Part of the restoration included the fitting of a Valenta engine as the original had long been removed. It was found, however, that fitting a standard Valenta was far from a simple process as there were many parts in the prototype that were different from the production Class 43s.

The driver's cab and controls of No. 41 001 at the Didcot Diesel Gala, 25 May 2014.

The slab-sided rear cab of prototype Class 41 No. 41 001 showing the gangway and, to the right of that, the windscreen of the auxiliary driving position (which had simplified controls). Also note the glazed door to the guard's accommodation and another window to the rear of that. Taken at the Didcot Diesel Gala, 25 May 2014.

Production sets were known as High Speed Trains (HSTs) rather than HSDTs and designated Class 253. Pictured at Bristol Temple Meads in October 1983 is the very first production power car built, No. W43002, of the first production set No. 253 001. The car is in BR Inter City 125 'rail blue' and 'rail grey' livery, with the set number in black on the coupling cover and the power car number in white on the guard's compartment. No. W43002 was later named *Top of the Pops* and subsequently renamed *Techniquest*. The production power cars had the same 2,250 hp Paxman 12RP200 Valenta engine as in the prototypes, but a completely revised cab by industrial designer Kenneth Granger. Catering for a crew of two, the windscreen was much larger, the lower windscreen housing the marker lights was replaced by two separate narrow-glazed covers over triple marker light clusters, side windows to the cab were fitted to improve sideways visibility, no buffers were fitted, and a more raked front replaced the rather blunt original. Modified and strengthened bogies were used. As for the rear guard's compartment, the luggage space was increased and the rear driving position was removed (the power cars no longer being expected to handle trains other than HSTs). However, the rear end window was retained as were the side windows behind the access door. The first 151 power cars were so built and were similarly classified as the prototypes – Driving Motor Brakes (DMBs). However, it quickly became apparent that conditions for the guard were unsatisfactory so it was decided to move his accommodation to one of the coaches instead, which also included some luggage facilities. The power car's rear compartment was then used solely for luggage and reclassified Driving Motors (DMs). Eventually, the rear end window and the two rear side windows were removed as superfluous.

At Bristol Temple Meads in May 1991 is Class 43 No. 43 193 *Yorkshire Post* (later *Plymouth City of Discovery* and then *Rio Triumph*) in BR Inter City 125 Executive livery. The power car has a large number in black on the cab front below the driver's window and small black numbers on the former guard's compartment at the rear. This locomotive was one of the last batch of forty-six to be built without rear end and side windows to the luggage compartment, the guard now being housed in a coach. This batch was classified as Driving Motors (DMs) from new.

Class 43/0 No. 43 045 (formerly *The Grammar School Doncaster AD 1350*) of East Midlands Trains in unbranded Midland Mainline livery at the head of an HST at Derby on a St Pancras service, 15 July 2008. No. 43 045 has been refurbished with a Paxman 12VP185 engine and triple light clusters. It was BR who first initiated a refurbishment and re-engining programme for the Class 43s, but the first trials with a 2,400 hp Mirlees Blackstone 12MB190 engine in 1986–87 proved abortive when increased sectorisation of BR resulted in the Mirlees programme being dropped in favour of the 2,100 hp Paxman 12VP185, trialled in 1994. Production refurbishments followed in 1995, although by that time privatisation had started so deliveries went to the new train operating companies – Great Western, Midland Mainline and Great North Eastern Railway. In the event, First Great Western and GNER chose to standardise on MTU engines instead but MMl and its successor East Midlands Trains decided to continue BR's VP185 re-engining programme until all their fleet were so equipped by 2003. At the same time, MMl and EMT replaced the standard marker light arrangement by individual triple light clusters. Moreover, some of the Valenta-engined locomotives were fitted with the new light clusters before re-engining. Although not officially re-classified from Class 43, when other train operators reclassified their locomotives Class 43/2, MMls effectively became Class 43/0. Several MMl/EMT locomotives that eventually proved surplus were taken over by GNER/National Express East Coast.

First Great Western, which inherited several Paxman VP185-engined Class 43s from BR, decided this engine did not meet its needs and instead adopted the German 2,250 hp MTU 16V4000R14R engine, at the same time undertaking an extensive re-engineering programme. This involved twin individual light clusters, in comparison to MMl's triple arrangement. The locomotives also received a new livery of plain indigo blue ('reflex blue'). The locomotives were not officially reclassified but effectively became Class 43/0 when the refurbished locomotives of other operators became Class 43/2. FGW's re-engining programme lasted from 2005 to 2009. The photograph shows Class 43/0 No. 43 073 *Penydarren* at Bristol Temple Meads in May 2010, showing off the unimaginative blandness of FGW's plain blue livery to perfection. Network Rail followed a similar refurbishment programme to FGW's and fitted MTU engines to its small fleet in 2009.

When the Great North Eastern Railway embarked on their refurbishment programme in 2006, they decided to follow FGW's example and adopt the highly efficient and reliable German MTU engine; the locomotives were then renumbered by adding 200 to their original number and the locos reclassified 43/2, though why GNER saw the need to reclassify when MMl and FGW didn't is unclear. The refurbishment was as extensive as FGW's although not identical, and completed by National Express East Coast in 2009. Similar twin light clusters to FGW's refurbished locomotives were fitted, and these were also fitted to some of GNER's Valenta-powered examples. Pictured here is Class 43/2 No. 43 367 (ex-No. 43 167) *Deltic 50 1955–2005* with a '10 per cent discount on line' advert of National Express East Coach in their interim livery at Kings Cross on an Aberdeen service, 13 December 2007. When NXEC took over the East Coast Main Line franchise from GNER, they used the same deep blue base livery, but replaced the red-orange stripe with a white one – but note a remnant of an orange stripe on the door of the first coach.

Virgin Cross Country took a completely different course to other HST operators and simply got rid of its fleet, replacing them entirely with new Class 220 Voyager and 221 Super Voyager DMUs. However, a large increase in passenger numbers which followed the DMUs' introduction resulted in Virgin's successor (Arriva) Cross Country deciding to reintroduce some HSTs to supplement the Voyagers inherited from Virgin. Initially, Arriva had to hire in HSTs from Midland Mainline and National Express East Coast but eventually obtained their own fleet out of storage. Arriva also went down the FGW route and refitted their Class 43s with MTU engines and twin light clusters in 2008–2009, reclassifying them Class 43/2s like GNER. Seen here is Class 43/2 No. 43 378 (ex-No. 43 178) of Cross Country in their full livery at Bristol Temple Meads on a Dundee–Newquay service, 22 August 2009.

GNER and its successor National Express East Coast Class took over a number of Midland Mainline locomotives that were surplus to MMI's requirements. These had triple light clusters and both the original Paxman Valenta engines and the later Paxman 12VP185 engines. However, these were soon re-engined with MTU engines which GNER/NXEC had standardised. Pictured here is a former MMI loco with triple light clusters which had a Valenta engine, but which GNER had replaced by a MTU engine. It is Class 43/2 No. 43 251 (previously No. 43 051 *The Duke and Duchess of York*), although with no number apparent. What is apparent is a Tutankhamun advertisement. At this time (13 December 2007) the ECML franchise had recently been taken over by National Express East Coast and the loco is in de-branded GNER livery with no indication of the new operator. The train is at Kings Cross on an Inverness service.

The buffered Class 43s were part of a stop-gap push-pull scheme when the ECML was electrified in the late 1980s; the construction of the Mark 4 coaches (including driver van trailers) were delayed so the newly completed Class 91 electric locos were perforce set to haul Mark 3 coaches with, at the other end, buffer-fitted Class 43s converted from standard examples, initially acting as DVTs but later with their engines providing normal power. The marker light clusters remained the standard two groups each under a glazed cover. Although these locos were returned to normal HST use after the MK4 sets were in service, by the mid-2000s all (except two used by Network Rail) were out of use and in a decrepit condition. Rescue came in the form of new Open Access Operator Grand Central which totally re-engineered the Class 43s, including the fitting of twin light clusters, but retained the Valenta engines. These became the last Valenta-powered Class 43s in service, but eventually Grand Central also went down the MTU route, re-engining their fleet in 2010, when they became Class 43/2. However, the picture shows a Grand Central Class 43 which is still Valenta-engined: No. 43 123 in GC's surprisingly attractive black livery arriving at Kings Cross on a Sunderland service, 30 June 2008. Note the damaged warning horn grill.

The BP10 power bogie of a Class 43 locomotive No.43 088 (ex-*Rio Campaigner*, ex-*XIII Commonwealth Games Scotland 1986*) of First Great Western at Bristol Temple Meads, 11 April 2014. Although a lightweight, fabricated design, they are one of the strongest bogie types ever used on British railways. The primary suspension employs coil springs with Alsthom links and the secondary suspension have Flexicoil springs and Koni dampers. The prototype bogies were tested on Class 86 electric locomotive No.E3173 in 1969.

Brush Works plate of Class 43 HST locomotive No. 43 142 *Reading Panel Signal Box 1965–2010* (ex-*St Mary's Hospital Paddington*). The plate indicates the locomotive's refurbishment, which included the fitting of a MTU engine. The photo was taken at Bristol Temple Meads, 11 April 2014.

RE-ENGINEERED BY
BRUSH
TRACTION
2007

Class 43 information panel and emblem of Old Oak Common MPD. At Paddington.

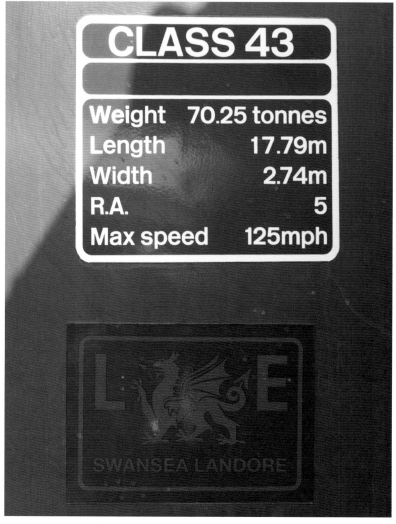

Class 43 Information panel and dragon emblem of Swansea (Landore) MPD. At Bristol Temple Meads, 11 April 2014.

HST MK.3 TRAILERS/COACHES –
A COMPARISON

TRFB (trailer restaurant first buffet) No. 40717 of First Great Western in their older indigo-blue and gold livery, built by BREL (Derby) in 1979 as Mk. 3 TRUB (trailer restaurant unclassified buffet) No. 40317 and later converted to a TRFB. At Bristol Temple Meads, 5 July 2008.

TRSB (trailer restaurant second buffet) No. 40424 of Grand Central in their black livery, built by BREL (Derby) in 1977 as HST Mk. 3 TRUB No. 40024 and later converted to a TRSB. Note knife and fork logo on the door – no mistaking what this coach is for! Photographed at Kings Cross, 30 June 2008.

TCC (trailer composite catering) No. 45001 of Cross Country at Bristol Temple Meads, 25 March 2009. Built by BREL (Derby) in 1975 as loco-hauled Mk. 3a TSO (trailer standard open) No. 12004 and rebuilt to a Mk.3 TCC in 2008.

TSMB (trailer standard miniature buffet) No. 40111 of First Great Western in their 'dynamic lines' livery. Built by BREL (Derby) 1979 as TS (trailer standard) No. 42248 and subsequently converted to a TSMB at Laira in 2010 by the installation of a miniature buffet. Note the blanked-out window and half-window behind which the miniature buffet is located. For some reason, the TSMBs have a yellow stripe (which normally denotes first class) rather than a red one (which normally denotes catering facilities) at the buffet end of the coach. Photographed at Bristol Temple Meads, 11 April 2014.

TF (trailer first) No. 41084 coach of East Midlands Trains still in Midland Mainline livery at Derby, 15 July 2008. It was built by BREL (Derby) in 1977.

TS (trailer standard) No. 42402 of Grand Central in their black livery, built by BREL (Derby) in 1977 as a loco-hauled Mk. 3a TSO (trailer standard open) No.12155 and converted to form part of an HST in 2008. At Kings Cross, 5 July 2008. Note the silver doors to signify standard class.

TGS (trailer guard standard) No. 44063 of National Express East Coast in their livery at Newcastle Central, 2 September 2008. The coach was built by BREL (Derby) in 1981.

TS(d) (trailer standard disabled) No. 42347 of First Great Western in their 'dynamic lines' livery, built by BREL (Derby) in 1977 as a Mk. 3 Trailer First No. 41054 and later converted to a TS(d) with a disabled toilet. At Bristol Temple Meads, 5 July 2008.

BR BT10 bogie of Mk. 3 HST coach TF (trailer first) No. 41015 of First Great Western at Bristol Temple Meads, 11 April 2012.

Interior of a Mk. 3 TF (trailer first) of First Great Western. The leather seats are Primarius and are as comfortable as they look. Pictured at Paddington, 1 April 2009.

Interior of a Mk. 3 TS (trailer standard) of First Great Western. The seating is in high density configuration and the seats are of Grammer – or is it 'Crammer'? – design. Pictured at Paddington, 7 May 2014.

BIBLIOGRAPHY

Marsden, Colin J., *HST Silver Jubilee* (Hersham, Ian Allen Publishing Ltd: 2001)

Marsden, Colin J., *HST The Second Millennium* (Hersham, Ian Allen Publishing Ltd: 2010)

Marsden, Colin J., *Modern Locomotives Illustrated Nos. 208 & 209 The HST* (Stamford, Key Publishing Ltd: 2014)

Marsden, Colin J., *The Power of the HSTs* (Oxford, OPC: 2006)

Morrison, Gavin, *The Heyday of the HST* (Hersham, Ian Allen Publishing Ltd: 2007)

ABBREVIATIONS

APT	Advanced Passenger Train
BR	British Rail
BREL	British Rail Engineering Limited
DMU	Diesel multiple unit
DVT	Driving Van Trailer
ECML	East Coast Main Line
EMT	East Midlands Trains
ER	Eastern Region
FGW	First Great Western
GC	Grand Central
GNER	Great North Eastern Railway
GWML	Great Western Main Line
HSDT	High Speed Diesel Train
HST	High Speed Train
MML	Midland Main Line (the route)
MMl	Midland Mainline (the train operating company that operated the MML franchise for several years)
NXEC	National Express East Coast
PC	Power Car
WR	Western Region
XC	Cross Country